PowerKids Readers:

ROAD MACHINES

Road Milling Machines

Joanne Randolph

The Rosen Publishing Group's
PowerKids Press™
New York

T0078481

1

For Joseph Hobson, with love

Published in 2002 by The Rosen Publishing Group, Inc.
29 East 21st Street, New York, NY 10010

Copyright © 2002 by The Rosen Publishing Group, Inc.

All rights reserved. No part of this book may be reproduced in any form without permission in writing from the publisher, except by a reviewer.

First Edition

Book Design: Michael Donnellan

Photo Credits: pp. 1, 13, 15, 17 © Highway Images/Genat; pp. 5, 7, 9, 11, 19, 21 by Beryl Goldberg.

Randolph, Joanne.
Road milling machines / Joanne Randolph.—1st ed.
    p. cm. — (Road machines)
Includes bibliographical references and index.
  ISBN 0–8239–6041–2 (library binding)
1. Cold planers-Juvenile literature. [1. Road machinery.] 1.Title.
  TE223 .R3598 2002
  625.8'5—dc21

                                                    2001000654

Manufactured in the United States of America

Contents

NORTHLAKE PUBLIC LIBRARY DIST.
231 NORTH WOLF ROAD
NORTHLAKE, ILLINOIS 60164

A road miller is a
big machine.

5

A miller helps make old roads new! A miller rips up the street. Then another machine can pave a new, smooth road.

⚠ WARNING

THIS MACHINE MAY FALL AND CAUSE
PERSONAL INJURY OR DEATH. ALWAYS
INSTALL SUPPORT LEGS BEFORE
WORKING BENEATH MACHINE. SEE
OPERATORS MANUAL FOR INSTRUCTION.

9

The miller has a wheel with huge teeth. The teeth rip up the street.

11

Grooves are left where the teeth have ripped up the road. The grooves look like dents. The new street will fill in these grooves and stick better.

15

A person drives the miller. He makes sure the miller does its job ripping up the street.

This man is fixing a miller's teeth. He makes sure the miller is working right.

19

Road milling machines help keep our streets in great shape!

Words to Know

conveyor

crawler tracks

grooves

road milling machine

Here are more books to read about road milling machines:
Diggers and other Construction Machines
by Jon Richards
Copper Beech Books

To learn more about road milling machines, check out this Web site:
www.field-guides.com/html/cold_planer.html

Index

Word Count: 163
Note to Librarians, Teachers, and Parents

PowerKids Readers are specially designed to help emergent and beginning readers build their skills in reading for information. Simple vocabulary and concepts are paired with photographs of real kids in real-life situations or stunning, detailed images from the natural world around them. Readers will respond to written language by linking meaning with their own everyday experiences and observations. Sentences are short and simple, employing a basic vocabulary of sight words, as well as new words that describe objects or processes that take place in the natural world. Large type, clean design, and photographs corresponding directly to the text all help children to decipher meaning. Features such as a contents page, picture glossary, and index help children get the most out of PowerKids Readers. They also introduce children to the basic elements of a book, which they will encounter in their future reading experiences. Lists of related books and Web sites encourage kids to explore other sources and to continue the process of learning.